Karoly Nyisztor

AGILE, SCRUM, AND KANBAN FOUNDATIONS

A PRACTICAL GUIDE TO AGILE SOFTWARE DEVELOPMENT

Copyright © 2024 Karoly Nyisztor

All rights reserved.

ISBN: 9798366791663

Contents

1. Getting Started — 1
1.1 Introduction — 2
1.2 Prerequisites — 3

2 Agile Software Development — 4
2.1 What is Agile? — 5
2.2 The Traditional Way of Developing Software — 7
2.3 Dealing with Uncertainty in Software Development — 9
2.4 Agile Origins: The Agile Manifesto — 11
2.5 What Does it Take to Become Agile? — 13

3 Scrum — 15
3.1 The Basics of Scrum — 16
3.2 The Product Owner — 18
3.3 Common Myths about the Product Owner Role — 20
3.4 The Scrum Master — 22
3.5 The Development Team — 24
3.6 Sprint Planning — 26
3.7 Planning Poker and the Psychology of Estimating — 28
3.8 The Daily Stand Up — 31
3.9 The Sprint Review — 33
3.10 How to Improve Your Sprint Review Meetings — 34
3.11 The Sprint Retrospective: Reflecting on the Past — 37
3.12 Documentation in Scrum - Misconceptions and Best Practices — 39

4. Kanban — 41
4.1 Introducing Kanban — 42
4.2 Setting up a Kanban Board in Asana — 45
4.3 The Importance of Limiting Work In Progress — 47
4.4 Creating a Cumulative Flow Diagram in Excel — 49

4.5 How to Interpret a CFD? 51

4.6 Goodbye! 56

1. GETTING STARTED

1.1 Introduction

Planning complex software projects, organizing the work, and tracking the progress aren't straightforward tasks. There are many moving parts, and to make matters worse, the team setup and the original goals may change over time.

How do we deal with this complexity and unpredictability?

Over the years, with the increasing need for software systems, this question popped up more and more frequently.

Software developers started to explore various practices, and gradually it became apparent what was working and what wasn't. The results were distilled into a set of principles and values that stand at the core of the Agile approach to software development.

Now, while Agile is a perfect fit for projects that can't be planned upfront—that is, in the vast majority of cases—successfully implementing it can be tricky.

It requires us to think differently when it comes to organizing our work. Taking on just some of its elements—for example, working in shorter timeframes—won't make us Agile.

To increase our productivity, we first need to understand the core values behind Agile practices. And that's precisely why I wrote this book. I will address many questions and help you avoid the costly mistakes many organizations make when trying to adopt Agile principles.

We'll then delve into Scrum, the most widespread Agile framework used to organize teams, set goals, administer work, and measure progress.

Next, I'll show you how to use Kanban techniques to manage work in progress and improve the efficiency of development teams.

Throughout this book, I'll share practical examples to solidify the presented concepts and techniques.

By the end, you'll know how to organize your work and deliver complex projects successfully while responding to changing business needs.

Are you ready to get started? Let's dive in!

1.2 Prerequisites

I designed this book so that it provides value even if you don't have a background in software development. That said, you'll find this material especially helpful if you're part of a development team trying to embrace the Agile mindset.

You'll appreciate this book even if your work is not directly related to coding because it provides essential clues about organizing work effectively and overcoming uncertainty. For example, I'm using a Kanban board to track my progress with this project—we'll talk about Kanban a bit later.

Anyway, the bottom line is that you don't need any prior experience in software development or anything else.

Throughout the book, I'll share insights and real-life stories from my own professional experience to support your understanding. I'll also provide various resources and quizzes to help you solidify these concepts.

We'll use some tools and apps that are freely available—although they may offer some premium features, you'll do just fine with the free options.

Now, with all this out of the way, let's continue our journey.

2 AGILE SOFTWARE DEVELOPMENT

2.1 What is Agile?

You might have heard about Agile in the context of software development. But what exactly does it mean? There seems to be a lot of confusion and misconceptions surrounding the term Agile. So let me try to explain it.

I'll first illustrate the difference between traditional software development and the agile approach using an example. Let's say you want to travel somewhere. The train will get you to where you want to go, but it's hard to change course.

Trains have fixed routes and schedules that make changes difficult. Severe weather conditions, obstacles on the tracks, and other unforeseen events can affect your travel.

However, if you travel by car, you can easily change course if something unexpected happens.

A traditional organization is like a train: the tracks are laid, everything is planned in advance, and you follow these tracks precisely to get to your destination—that is, to finish the project. However, this model is not prepared to respond to change.

Agile organizations are more like a car: you can easily adjust your travel plans and routes whenever necessary.

Changing circumstances also affect software projects: even when a project has been approved, the original goals or the underlying technology might change. The chance of this happening increases sharply with the complexity of the software project. That's because complexity translates to unclear requirements and longer development times. This flexibility doesn't come for free, though. Like our car analogy, making abrupt changes means longer travel times, higher costs—and these are also costs associated with the agile approach that we need to take into account..

Story Time

Now, let me tell you a real story—a situation where agility could have saved a lot of time, energy, and frustration.

I once worked on a project that relied on a new set of tools. The idea was promising and quite revolutionary back then. In theory, one could build apps using a mouse by connecting UI elements with pre-defined logic. We were the lucky team to try out this new tool.

We were given three weeks to finish the project. Everybody was happy.

What could go wrong, right? The idea behind these tools came from the same visionary guy who was behind so many other innovative ideas.

However, shortly after starting the project, we realized that the technology wasn't ready for prime time. The tools were buggy and inefficient. The IDE kept crashing. We spent most of our time rebooting our computers and checking the system logs for errors. When we finally managed to implement—sorry, drag and drop—a feature, the software would literally break apart. What worked on one machine refused to start up on another. It was nerve-wracking!

We had no clue about the cause of these unexpected issues because there was no actual code we wrote. We were basically working with black boxes, and there was no way to find out what was going on after connecting the UI elements with the logic.

We kept reporting the issues. Management wasn't happy, to say the least. "It has to work!", "You just need to put in more work," "Perhaps you need more RAM or a faster CPU"—these were the usual responses to our complaints.

However, after two-and-a-half weeks, it became clear we wouldn't be able to make it work. The team was frustrated, and we felt guilty, despite knowing full well it wasn't our fault. We were testing a toolset that was released for production prematurely.

An agile approach would have surfaced the problems earlier, giving us a chance to stop spending our time on a lost cause—or go ahead using a different, more reliable technology stack.

By the way, the company made a few more attempts to use the "innovative" drag-and-drop approach to develop actual products. All attempts ended in failure, and the toolset was eventually abandoned.

2.2 The Traditional Way of Developing Software

Software development is inherently different from building cars or houses, or even computer hardware. But early software development tried to follow the approach used in construction and manufacturing projects.

The main idea behind this approach is that you need to have a detailed plan, comprehensive documentation, and established processes and rules that workers need to follow. This approach makes total sense when building a house or a car: a detailed blueprint must be defined in advance, and we must also know what tools and processes will be used before starting any actual work.

And from the moment we start building the house or manufacturing the cars, the plans are pretty much locked in. Making changes as you go along can have disastrous effects on the initial budget and deadline.

The software industry adopted this approach and named it the Waterfall Model. The Waterfall model requires a plan to be created and approved before any work on the product begins. This helps to mitigate risk by providing a detailed estimation of cost and time to completion.

The Waterfall Model has five distinct phases:

1. Requirements
2. Analysis and Design
3. Development
4. Testing and
5. Deployment and Maintenance

Progress flows steadily downwards (like a waterfall) through these discrete phases.

Each phase produces an output that's used by the following. Thus, a phase cannot start until the previous one is completed. The name is intended to convey similarity with the way water flows over a cliff (a waterfall) from a higher level to a lower one.

The Requirements phase captures the requirements, which in turn are used as input by the Analysis and Design phase. The Analysis and Design phase produces a comprehensive system design and detailed documentation. Development can only start when these documents are complete. Software developers need to follow the design documentation to implement the product.

When the development phase is completed, we start testing and fixing bugs, and once the software is ready, it gets released. The maintenance phase involves fixing bugs that were not caught during the testing phase and adding new features requested by customers.

This seems like a pretty reasonable approach to software development. Additionally, the Waterfall model is based on battle-proven engineering principles that underlie the manufacturing process.

So, why do we need alternatives like Agile development?

Well, Waterfall is not without drawbacks. While it works nicely for certain projects, it comes with two flaws you need to consider.

First, the customer doesn't have a chance to check out the product until the end of the testing or, even worse, the deployment phase. What if we misunderstood the customer and built a product nobody wanted? We'll only find it out after spending a lot of time and money to build the wrong product.

This leads to the second major problem with this model: it assumes that requirements won't change over time. Unfortunately, that's not always the case. Unlike in car manufacturing, the need for constantly refining a product to fit customer needs can happen as the project moves forward. However, the Waterfall model locks in the requirements at the beginning, with no or almost no way to change them later. Recreating a plan involves significant costs should it need changing or be unsuitable once implemented.

Having said that, Waterfall has its place and works well when the initial requirements don't change often or abruptly.

But how do we handle less-predictable situations? I'll answer this question next.

2.3 Dealing with Uncertainty in Software Development

We've seen that one of Waterfall's biggest weaknesses is its inability to respond to change. The other problem is that it could take several months or even years before customers could see the results of their investment.

Bottom line: the methods borrowed from traditional manufacturing don't always work in software projects.

Now, let's have a look at how Agile is different and why it offers a better solution in uncertain environments.

We'll start with the first problem: *lack of responsiveness to change.*

Instead of trying to predict everything and making all the important decisions ahead of time, Agile works in short iterations. The requirements are set, and functionality is developed and delivered periodically in small increments. This incremental approach makes it possible to review functionality early on and often. Agile relies on collaboration between stakeholders to decide what is needed next. Unlike the more rigid, plan-driven Waterfall model, not having fixed, pre-defined requirements makes it easier to respond to changes.

The other issue with traditional software development is that *customers have to wait* until the very end to try out the product.

Agile avoids this problem by delivering working software frequently. At the end of each iteration, the software is tested and evaluated by the customer. The product definition and the requirements may change based on client feedback, allowing constant course corrections to fit needs as they arise.

The opportunity to test and evaluate the product while still in development ensures that we're building the right thing while preventing costly changes.

There have been countless cases when time and money were spent developing features only to discover that no one needed them.

Do you remember Clippy, the animated, talking paper clip that was supposed to provide you with guidance when working in Microsoft Office?

Clippy first appeared in Office 97, and it became instantly unpopular. The assistant was often wrong and annoying and never really gained traction. Yet it took about ten years until it was completely removed from Microsoft Office.

Agile's incremental approach would have revealed that users didn't want an assistant that popped up every time they typed, and Clippy would never have made it to production.

All that said, Agile is no silver bullet, but it does allow us to react quickly to feedback and change requirements early on without wasting more money.

Agile is a better choice than Waterfall in cases where the definition of the final product and its requirements are subject to change. On the other hand, Waterfall works well when requirements do not vary significantly during the lifetime of the project—that is, when customers determine what they need upfront and if change requests are rare.

Next, we'll have a look at what's needed for Agile to work effectively.

2.4 Agile Origins: The Agile Manifesto

In 2001, a group of software developers nailed down the first official definition of the agile approach in a document called the "Agile Manifesto." This document defined four values that would become the cornerstones of agile development.

The four values in the Agile Manifesto are:

1. Individuals and Interactions Over Processes and Tools

The manifesto's first value stresses that people are at the forefront of the agile process instead of rigid rules or processes. The agile approach provides individuals with the opportunity to take an active role in developing software, encouraging discussion and collaboration along the way. Agile developers put their focus on creating products that meet the needs of real users rather than getting bogged down in the details of a particular technology or methodology.

The key to this value is face-to-face interactions—as opposed to communicating solely via email or phone. Agile developers work in small teams of about five to ten people who usually share an open office space.

This setup allows them to closely collaborate and fosters a comfortable work environment where feedback is easily exchanged.

As a side note: Agile teams are encouraged to separate their workspace by activities rather than individuals. (e.g., the coding team has a specific open space and the testing team has another one).

Also, specific tasks require focus and solitude. Open collaborative spaces can be detrimental in such cases.

Whenever I had to work on complex tasks, I booked a small meeting room where I could work in peace. Every team member knew where I was and what I was working on, so they wouldn't disturb me unless it was *really* needed. The practice of booking a private meeting room for deep work became a habit and part of my agile workflow.

My point here is that although communication is key to agile, it must not be mistaken for constant social interaction.

2. Working Software Over Comprehensive Documentation

The second value in the Agile Manifesto stresses that delivering a product should be a higher priority than creating extensive documentation. Since requirements can change quickly, it's not practical to write extensive documents ahead of time about what features a product will have and how they will be implemented.

Now, does that mean that Agile teams don't ever create documentation? Not at all! Agile teams do produce documentation, but only when it's useful and necessary. And when I say documentation, I don't mean Word or PDF files: developers often create visual mockups to communicate with customers or insert comments directly into code to capture technical details.

3. Customer Collaboration Over Contract Negotiation

The third value states that working closely with the customers should be a priority over contract negotiation. By talking directly to customers about what features they need and how to implement those features, agile developers discover more about their customers' needs than any contract or specification could possibly express.

Agile teams don't spend time negotiating requirements with customers as if they were lawyers preparing a contract. If the customer becomes unhappy with the product, the developer can talk directly with the customer again to find out why and discuss whether any changes are possible.

4. Responding to Change over Following a Plan

The fourth value states that responding to change should be a priority over following a plan. That doesn't mean that agile developers don't plan before starting to build a specific product or feature. They do, but they also accept that their plans may change frequently. To minimize the risks, agile developers plan in short iterations.

To sum up: Agile is a software development approach based on the Agile Manifesto, which defines how to deliver high-quality, cost-effective, and flexible software systems effectively.

2.5 What Does it Take to Become Agile?

Even if the benefits of Agile are clear, transitioning to Agile is easier said than done.

Let's take a look at some of the most common challenges that organizations face when adopting Agile, as well as steps you can take to avoid them.

One of the most common mistakes is when companies **start with a big planning phase for their Agile transition before even trying to sell the idea internally.**

I can't forget how frustrated we were when our company announced that we were going to transition from Waterfall to Agile. We had no idea what Agile meant, so we started to research the topic on our own. The results were either the official Scrum documentation or horror stories from people who tried to transition and failed.

The company hired an Agile expert to explain the process.

He kept emphasizing how Agile is necessary for us to do effective software development. But he was unaware of the problems we had been facing before. So, he was unable to convince us of the benefits of Agile.

At one point, he even stated: "I've never seen any team **not** fail at the beginning, but they get better over time." He meant this to be funny, but nobody smiled. We were already under a lot of pressure, and what we needed the least was even more uncertainty.

Looking back, a better approach would have been to start with a pilot team and allow this first group of people to experiment with Agile first-hand. This would have provided us with relevant information about whether or not Agile makes sense for our organization.

Another big challenge is that Agile assumes a willingness to change the company's structure and culture.

Traditional organizations use formal hierarchies and processes that involve a lot of direct management and vertical communication.

In Agile, the teams need to have more autonomy. They are responsible for their own work from beginning to end. Agile emphasizes transparency, collaboration, and consensus within teams to get the best work done through rapid iteration.

Changing a hierarchical, top-down structure to a flatter team structure brings about structural challenges for many companies. These changes can be difficult to achieve without causing chaos and unrest among employees.

Luckily there are approaches to Agile adoption that mitigate these challenges. A great example is what we call a "change agent," a role designed to help the organization transition to Agile and maintain its new behaviors even after they've been adopted.

People who are successful in this role understand the organization's business processes, objectives, and culture from the very start. They are able to draw on this knowledge when working with the organization's teams and management. A change agent can help provide insights into how Agile can fit into existing processes, as well as advise managers on how best to coach their teams during the transition period.

Another part of an effective Agile adoption strategy is establishing the so-called "Servant Leader." This term was coined by Robert Greenleaf in 1970. The idea behind it is that leadership should focus on guiding and empowering employees to achieve a common goal.

"Servant leadership" might sound like a management buzzword, but implemented the right way, it can actually be very effective at building trust between employees and managers.

A servant leader helps people reach their full potential while leading by example in an effort to build trust within the organization. This is important for Agile adoption because team members need to know that management has complete confidence in their decision-making process.

Switching to Agile comes with a number of challenges. The most common mistakes are rushing your adoption strategy and not considering change management during implementation. If you avoid these, you'll be well on your way to building a sustainable Agile organization.

3 SCRUM

3.1 The Basics of Scrum

So far, we've talked about the principles and values of Agile. While these principles and values are essential, they cannot be directly translated into a concrete methodology.

This is where Scrum comes in. The term Scrum was first introduced by Jeff Sutherland and Ken Schwaber in 1995. Scrum takes the values, and the principles defined in the Agile Manifesto and distills them into a simple-to-follow framework. Scrum provides general guidance for managing complex product development, but it does not define detailed actions. Instead, it focuses on a team-driven, empirical process.

How does this look in practice? Let's say you're working on a new product. The first step is to come up with an idea for the product. This can be done by talking to potential customers, doing market research, or simply coming up with a good idea on your own.

The next step is to define what the product should do. This is called the product backlog, which is a list of all the features and functionality that need to be included in the product. The product backlog is maintained and prioritized by the Product Owner, who's responsible for ensuring that the product meets the needs of the customer. In addition to the Product Owner, there are two other key roles in Scrum: Development Team and the Scrum Master.

The Development Team is responsible for building the product. They have all the skills needed to build a working product, including analysis and design.

The Scrum Master helps the team apply Scrum and follow its processes. The Scrum Master makes sure that the team has all it needs to be as productive as possible.

Now let's talk about how work is organized in Scrum.

Scrum uses fine-grained time boxes called Sprints. A Sprint is a fixed amount of time—typically 2-4 weeks long—during which specific work is done.

Before the start of each sprint, there is a planning meeting where the Product Owner and the Development Team members decide which Product Backlog items to select for the given sprint.

The Development Team analyzes the high-level items and converts them into a detailed list of tasks. The selected Product Backlog items and the detailed task list get merged into the Sprint Backlog document.

The Development Team is responsible for implementing these tasks and delivering the corresponding functionality. Each team member picks a task from the Sprint Backlog and starts working on it.

Development Team members then meet each day to assess their progress. The "Daily Scrum" (also called "Daily Standup") is a 15-minute meeting where the Development Team discusses what they've been working on, any obstacles that have surfaced, and what is going to be worked on next.

The Scrum Master and the Product Owner work with the team to remove any roadblocks or impediments. If the sprint is slipping behind, tasks are re-estimated and reprioritized.

Every sprint should result in a potentially shippable product increment—meaning that the product contains new features to show to customers.

At the end of each sprint, there is a Sprint Review Meeting for the Development Team to showcase what they have built.

This usually happens in the form of a product demo. The customers and other stakeholders get a chance to check out the new features and provide feedback. Direct feedback from customers is valuable input into the product development process.

The Sprint Review Meeting is followed by a Retrospective Meeting that goes over what was done during the sprint and what could be improved.

This meeting is not about fingerpointing but instead about identifying issues and coming up with constructive solutions.

Then, the process starts over again until the project completes.

Now that you've got an overview of Scrum, let's delve deeper into Scrum roles. We'll begin with the Product Owner.

3.2 The Product Owner

The Product Owner is an independent and self-managing position and it's also one of the most difficult on a scrum team. A good product owner makes the scrum team shine, while a bad one can hold back even the most experienced development team.

The Product Owner has two main responsibilities:

1. Represent the customer's interests
2. Facilitate conversations within the team

Let's start with the first responsibility. **The Product Owner represents the customer**, which translates into understanding their pain, their needs, and what they hope to get out of using your product. This is an ideal situation because it ensures that the product you are building actually solves a need in the market.

The second responsibility, **facilitating conversations within the team**, requires an understanding of what developers can actually build, considering their skills and technology constraints. More specifically, it requires an understanding of what they are capable of, how fast they can build something, and what is feasible with the available hardware and software stack. Therefore, a good Product Owner needs to be able to communicate effectively and validate ideas with the team.

A typical day for a Product Owner likely plays out as follows:

- → Get in touch with the team to discuss their current progress
- → Gain additional insight into customer feedback, see if there are any new requirements for the product
- → Seek clarification on requirements where necessary
- → Communicate priorities to the team, highlight key items that should be worked on first
- → Work with stakeholders in the company to manage customer expectations
- → Keep an eye out for key risks and dependencies that may arise, communicate these to the team
- → Continually seek ways of improving themselves and their work. Many Product Owners do this by trying to contribute towards writing user stories or acceptance criteria
- → Communicate progress on a regular basis with the team and stakeholders

Traditionally, the role of Product Owner was filled by someone much higher up in the organizational structure. They would consult with stakeholders and customers before creating requirements for what needed to be built. However, as more teams adopted agile, it became apparent that those closer to the product could provide more valuable input.

So, who can fill the Product Owner role?

Ideally, the Product Owner role will have intimate knowledge about the product and a good idea of what it needs to function. While you wouldn't expect a Product Owner to have a deep technical background, it is beneficial if they can speak the language of those who do. This makes it easier for them to communicate the requirements with those implementing them.

The Product Owner defines what "done" means to ensure that what is built meets the needs of your clients. Now, "done" could mean many different things. For example, one team I worked with was building a framework that provided straightforward, fast, and secure access to the company's web services. Their "done" criteria were not just signing off on the project but also completing it in a way that allowed developers to build reliable apps effectively while avoiding security risks and performance issues.

3.3 Common Myths about the Product Owner Role

The Scrum Product Owner role is often misunderstood or misinterpreted.

One of the most common myths is that the Product Owner is a product or project manager.

This couldn't be further from the truth. The Product Owner and the Project Manager are typically not even on the same organizational chart.

The Product Owner is not a proxy for the customer either. There was an "old school" approach to the implementation of Scrum, which included the customer—or someone with authority to make decisions on their behalf, such as a manager—in the Scrum team. This person was directly involved in all aspects of developing requirements and making decisions.

It was believed that by including the customer in the team, the Scrum team would receive direct orders from the customer, which would mean less management overhead.

This approach was doomed to failure. Here's why.

Customers typically lacked most—if not all—of the knowledge necessary to make optimal decisions. They suffered from analysis paralysis and usually delayed the delivery of a solution again and again. This made their involvement in the development process counterproductive.

The Scrum team made its own assumptions about what the customer really needed and preferred. Depending on the customer's behavior, this also had negative effects on the development process.

It became painfully obvious that without someone available to make decisions on behalf of the team, progress was stagnant, and the result was a lot of frustration.

The Product Owner is also not a proxy for the development team. This interpretation of the Product Owner role often leads to confusion and conflict as to who makes decisions on what can be delivered by the Scrum Team. The team should be self-organizing and make their own decisions as to how they will develop the features and functionality of the product.

The Product Owner's focus should be on the business side. They must balance process knowledge and business knowledge. This is the only way to make sure that the team will receive what it needs to achieve its goals.

The Product Owner is not accountable for costs or schedules. The Product

Owner's job is to maximize the value of the delivery while maintaining a healthy balance of scope, cost, and schedule, but the Product Owner is not directly accountable for these items.

The most common question I hear from new Product Owners is, *"Who's empowered to make a decision?"* Here's the simple answer: The Scrum Team needs consensus on what can be built within scope, cost, and schedule limits - with answers coming from self organization. The Product Owner facilitates this discussion, but decisions should be made by the Scrum Team. This is key to making Scrum work and helps avoid the common trap of "requirements churn."

The Product Owner and the Scrum Master work closely together to ensure that the product is being built in a way that satisfies the needs of their customer. However, it is important for them not to have overlapping responsibilities since they each serve different functions within the Scrum framework.

We'll talk about the Scrum Master role next.

3.4 The Scrum Master

The Scrum Master, in essence, ensures Scrum works—this is probably the shortest definition of this role.

The name Scrum Master can be confusing. It might give us the impression that there is a master who leads or controls the Scrum team. However, this role is closer to being a trainer, facilitator, and guardian.

The concept of an agile coach can be traced back to the Agile Manifesto. As a **trainer**, the Scrum Master makes sure that developers understand the real agile values and coaches them when needed with respect to the agile techniques used in Scrum.

When acting as a **facilitator**, the Scrum Master removes impediments that stand in the way of the Development Team—such as dealing with expiring software licenses, ordering new hardware, or organizing travel—in order to keep their focus on delivering working software. The Scrum Master ensures that all Scrum events take place and are positive, productive, and only as long as necessary (none of these events take longer than needed).

And as a **guardian**, they protect the team from external influences, interruptions, and distractions, which can interfere with the tasks at hand.

The Scrum Master's job is to create an environment where the team members are able to perform at their best so that they can deliver the maximum amount of business value in a sprint.

But there are a few myths surrounding the role. Let's now spend some time clarifying the role of the Scrum Master.

The Scrum Master **has no authority** over the team but does have influence which they use to be an effective servant leader. A servant leader should treat everyone with respect and dignity, seek the team's well-being, and trust that the team knows what they're doing.

The Scrum Master is **not responsible for estimating work**. However, they need to understand their team's velocity to know how much progress they can make in a sprint. They also need to understand their team's dynamics and any impediments that might prevent them from reaching their sprint goal. The Scrum Master should also identify any technical debt that could have an impact on the project and if any decisions are being blocked.

And while it is acceptable for a Scrum Master to participate in the development work—and be involved in coding, testing, and design—their **primary focus should be on facilitating the scrum process**.

To Conclude

The Scrum Master's role is unique, and nothing similar exists in traditional project management. Scrum Teams need a servant leader who understands the Scrum framework and helps hold them accountable. The Scrum Master is responsible for removing roadblocks, teaching Scrum concepts, and helping people with problems.

3.5 The Development Team

A development team is a group of people responsible for creating any aspect of a usable increment during each sprint. By usable increment, I mean anything that can be demonstrated to the product owner and stakeholders that adds value to the product. This could be completed features, a well-designed interface, or code that meets specific standards.

Tasks get assigned during sprint planning—I'll show you a real-life example of sprint planning in action later on.

Development teams should be small, between three and nine people, which allows for quick decision making and close collaboration. Bigger teams become less efficient, and fewer than three people aren't able to get enough work done in a sprint.

Smaller team sizes also ensure that the team can be fed with two pizzas over lunch.

The two-pizza rule was introduced by Jeff Bezos in the early days of Amazon. He said that each internal team should be small enough that it can be fed with two pizzas. And instituting this rule had nothing to do with the catering budget but rather the team's efficiency.

A team consists of software developers in most cases, but it may also include designers, architects, and other experts as needed. The composition of the team will vary depending on the product being built. For instance, we need different skill sets for a mobile app than we do for building an Enterprise Resource Planning system.

As a rule of thumb, the team should have all the skills necessary to take a product from ideation to delivery.

Another essential thing to remember is that a Scrum development team is a self-organizing unit. This means that they are responsible for deciding how best to complete the work assigned to them.

Of course, the product owner and stakeholders may provide input and guidance, but ultimately it is up to the team to determine the most efficient and effective way to get the job done.

It goes without saying that the development team must be able to work well together and communicate clearly. The team members need to be able to rely on each other and trust that everyone is doing their fair share. Otherwise, the project is likely to run into problems.

A common mistake when introducing Scrum is to try to create the "perfect" team. That's usually done by hand-picking its members and ensuring that everyone is a superstar.

While having a team of skilled individuals sounds great on paper, it doesn't always work out in practice. After all, even the best coders in the world will struggle if they can't communicate and collaborate effectively.

In my experience, it is better to start with a group of people who are willing to work together and then let them grow and develop into an effective team. They will inevitably make mistakes and run into problems, but that is part of the learning process.

Keep this in mind as you start to assemble your own development team. It's better to have a group of average performers who work well together than a group of A-players who can't stand each other.

Also, do not expect overnight results. Teams are organic entities that need time to gel.

Story Time

When I was first introduced to Scrum, I was working as a software developer at a company that was just starting to adopt the methodology. Our team was composed of coders who were used to working according to well-defined processes and procedures. We all knew each other and got along well enough.

However, the idea of self-organizing teams was foreign to us, and we were skeptical that it would work. The transition to Scrum came as a shock to us. We suddenly had to start attending daily standup meetings, estimating our work, and planning sprints. It was a lot of change all at once, and we struggled to adjust.

Fortunately, we had a good Scrum Master who was patient and helped us through the process. He explained the reasons behind the practices and taught us how to apply them to our work. Slowly but surely, we began to see the benefits of Scrum. So, don't be discouraged if your team experiences some growing pains in the beginning. It's normal and to be expected.

Now that we've talked about the development team, let's move on to sprint planning.

3.6 Sprint Planning

Sprint planning is where the Scrum team, including the Product Owner, discusses the prioritized backlog of work—known as product backlog items—and decides which ones to complete in the upcoming sprint. Each product backlog item is written as a user story, which helps to clearly communicate the desired functionality from the customer's perspective.

For example, a user story may be, "As a customer, I want to be able to track my order so that I know when it will arrive."

While anyone on the Scrum team can write user stories, the Product Owner typically gathers input from stakeholders, analyzes customer needs, and prioritizes user stories for maximum impact.

Effective user stories result in a clear understanding of project goals and help Scrum teams stay focused on delivering value to end users.

A user story would follow this template:

> "As a (type of user), I want to (desired action) so that (benefit)."

Now, for the "type of user" part, be specific. Instead of just saying "a user," try to identify a particular type, such as "a student" or "a marketing manager," because it really helps the team understand who they are building for and why.

To illustrate my point, here's a bad user story: "As a user, I want to see ads so that the company can make money."

That doesn't sound very natural or user-focused. Instead, try something like:

"As a marketing manager, I want to see targeted ads on the website so that we can increase conversions and drive revenue."

Much better, right? With a bit of refinement, the user story became clearer and more impactful.

While I can't give you a precise recipe for writing the perfect user story, remember to keep them concise and focused on the end user's needs.

Okay, back to sprint planning. Once the Scrum team commits to completing a set of user stories during the sprint, they break them down into tasks and estimate each one's level of effort—often measured in "story points."

Note that story points are not a direct representation of time. Instead, they're used to help estimate the amount of effort required to complete a user story. A story point value of "40" may not mean forty hours' worth of work—it depends on the agreement and understanding within the Scrum team. For instance, it may represent a task that the Scrum team agrees is about as difficult as any other task they've completed in past sprints that had a story point value of "40."

How do you convert story points to time? That's up to the Scrum team to decide.

Our team used an estimation strategy called "planning poker," where Scrum team members held up cards with numbers representing their estimate in hours for a story's level of effort. The next lecture will cover this technique in greater detail.

Alright, Scrum sprint planning—check. Next up, we'll dig into estimation using planning poker. Stay tuned!

3.7 Planning Poker and the Psychology of Estimating

I'd like to take a small detour here to talk about estimations. When we're planning our work, we need to estimate how long each task will take and decide how much work we can realistically accomplish in a sprint.

The problem is that estimates are notoriously difficult to get right. Now, I want to show you a technique that can help make your estimates more accurate. It's called Planning Poker.

The game goes something like this:

Each player—that is, Scrum team member—is given a deck of cards, each with a different number on it. The numbers represent the time in hours required to complete the task. Some cards use the Fibonacci sequence, which is a series of numbers where each number is the sum of the two previous numbers (1, 1, 2, 3, 5, 8, 13, 21, 34, 55, 89), but there are other variations as well.

Players look at the task and secretly choose the card that they think represents the amount of time required to complete the task.

Once everyone has chosen a card, all the players reveal their cards at the same time. If most of the players have chosen the same card, then that's the estimate.

Otherwise, the players with the highest and lowest estimates get a chance to explain their reasoning. After that, the process is repeated until there is consensus.

The Product Owner or the Scrum Master can use an egg timer to keep track of the time and make sure that the discussion doesn't go on for too long. When the timer runs out, the team needs to stop discussing, and a new round of poker begins.

Planning poker is a fun way of making estimates. I truly believe that it's a helpful tool for Agile sprint planning.

Now, whether we use planning poker or not, there'll always be some uncertainty when we're making estimates. And that's OK! Teams that have been using Scrum for a while usually have a good understanding of their capacity and are able to come up with more accurate estimates. Less experienced teams often overestimate the amount of work they can do.

It's human nature to be optimistic about our ability to get things done. And there might also be a desire to please the product owner or impress our colleagues by accepting more work than we can realistically complete. Some product owners may try to take advantage of this by adding more and more items to the sprint.

If we commit to too much work, we're setting ourselves up for failure. We end up working long hours and weekends, and the quality of our work suffers.

This is why it's important for the development team to have a clear understanding of their capacity and to be honest about what they can realistically complete in a sprint.

The product owner and development team should also keep in mind that the goal of sprint planning is not to fill up the entire sprint with work.

Here's my personal rule of thumb:

> The sprint should be no more than 75% full. This gives the team some breathing room and leaves room for unforeseen events. And unforeseen events always arise.

Story Time

Let me tell you a story. I've worked with a team of experienced developers who consistently delivered high-quality work on time.

One day, our product owner announced that he was leaving the company and his replacement would be starting the following week. The new product owner was a new hire who didn't have any experience with Scrum or Agile. He was very excited about the product and started adding new features to the backlog at an alarming rate.

The next sprint planning meeting was a disaster.

The product owner wanted to include so many new features that the sprint would have been more than 100% full. When we raised concerns about our ability to complete all the work, he assured us that we could handle it. He argued that we could easily take on more work. After all, we had always completed our sprints with time to spare.

We tried to explain that the extra capacity was there to allow for unforeseen events. But he didn't want to hear it. He wanted us to commit to more work.

As you might expect, the next few sprints didn't go well. We ended up working long hours and weekends. The quality of our work and the mood of the team suffered. And we were all very stressed out.

I don't know the exact reason, but he disappeared as quickly as he arrived. And we never heard from him again. His replacement was much more reasonable, and we were able to get back on track.

This story illustrates why all involved in sprint planning need to be realistic about what can be accomplished in a sprint. The development team should have a clear understanding of their capacity and be honest about what they can realistically complete.

And the product owner should remember that the goal is not to fill up the entire sprint with work. A consensus should be reached between the product owner and development team on what can be realistically achieved.

A sprint is not a race to see how much work can be done. It's a time-boxed period where we focus on delivering value to the customer.

3.8 The Daily Stand Up

The daily stand-up—also known as daily SCRUM—is a short (usually 15-minute long) meeting where the development team members meet to discuss the progress of the current sprint.

During the daily stand-up, each team member answers three questions:

1. What did you do yesterday?
2. What will you do today?
3. Are there any blockers to report? (drop)

These are the only questions that should be answered. If there are other questions that need to be discussed, then they should be saved for another time. The daily stand-up is not the time for long, drawn-out discussions. The goal is to keep the meeting focused on progress and blockers.

Here's an example of how one team member might respond to the three questions:

→ "Yesterday, I worked on the login page. I almost finished the front-end code, but I need to do some testing before it's done. "

→ "Today, I'm going to finish the login page and start working on the sign-up page."

→ "The only impediment is that I'm waiting on the design for the sign-up page. I'll follow up with the designer today."

You can clearly identify what was done yesterday, what needs to be done today, and any blockers.

The daily stand-up is not a time for solving problems. If there's an impediment, then it should be noted, and someone should be assigned to follow up on it. But the actual problem solving should happen after the stand-up.

The daily stand-up is a great way to keep the team focused and on track. It's also a good way to identify any problems early on so that they can be addressed quickly.

I'd like to share a few psychological tricks that can help keep the meeting short, focused, and effective.

Have Everyone Stand Up

It's helpful to have everyone stand up during the meeting. After all, it's called a stand-up for a reason. Standing up has been shown to improve focus and keep meetings shorter.

The Scrum Ball

We used a Scrum ball in our daily stand-ups that was passed around during the meeting. It doesn't have to be an actual ball. It can be any object that can be safely passed around.

The idea is that whoever is holding the ball is the only one who can talk. After answering the three questions, they pass the ball randomly to someone else in the room. This randomness has a beneficial effect: since people don't know when they will be called on, they tend to pay more attention during the meeting.

Keep It Short

To avoid getting sidetracked, it's important to keep the answers short and to the point.

We had a rule that if anyone went way over their time limit, they had to buy the whole team ice cream. It sounds silly, but it worked.

Same Place, Same Time

The last thing I want to mention about the daily stand-up is that it's crucial to have it at the same time and place every day. This might seem like a small detail, but it's actually quite important.

If the meeting is constantly being moved around or rescheduled, people will start to get frustrated, some will be late, and the meeting will become less effective.

3.9 The Sprint Review

At the end of each sprint, there is a sprint review meeting. This is where the development team presents the work they've completed to the product owner and stakeholders.

It usually goes like this:

First, the team chooses a presenter. Sometimes it takes two or more people to present the work, depending on how much there is.

Each team member should be given an opportunity to present at least once every few sprints. Unfortunately, in many organizations, the same people end up presenting all the time. While not everyone is a rock star presenter, it's important to give everyone a chance. This is a team effort, after all.

The presenter starts by giving an overview of the sprint and what was accomplished.

I'll stop right here to share a tip: avoid using PowerPoint or any other presentation software. Slides are often used as a crutch, and they tend to make sprint reviews longer and more boring. If you absolutely must use slides, then keep them to a minimum. Focusing on the actual work instead of boring everyone to death with a slide deck is a much better use of everyone's time. This reminds me of the following quote by Steve Jobs:

"People who know what they're talking about don't need PowerPoint."

So, try to briefly summarize the achievements of the sprint and move on to the demo.

The demo is where things get interesting. Technical demos—when presented well—can be really exciting and engaging. They can also be a great way to get feedback from the product owner and stakeholders. But all too often, they are dry and boring.

I've participated in many sprint review meetings, and I've seen them go both ways. When done right, they can be a great way to get everyone on the same page and collect feedback. But when done wrong, they can be a complete disaster.

In the next section, I'll share a few rules that will help you make sure your sprint review meetings are effective and engaging.

3.10 How to Improve Your Sprint Review Meetings

Do you dread your sprint review meetings? If so, you're not alone. But it doesn't have to be this way.

In this lecture, I'll share a few tips that will help make your sprint review meetings more effective and engaging.

1. Make the Demo Interesting

This may seem like a no-brainer, but you'd be surprised at how often sprint reviews turn into a snooze-fest.

To avoid this, the presenter should focus on the features that were completed during the sprint and how they work. Instead of going through each task and bug fix, it's better to focus on the overall functionality. A good demo should be short, sweet, and to the point.

2. Make the Demo Relevant to the Audience

Not everyone in the room is a technical expert. The demo should be pitched at the right level and avoid using overly technical terms.

I can't count the number of times I've sat through a demo where the presenter started talking about "polymorphic behavior" or "asynchronous execution" and lost almost everyone in the room. While technical terms are fine—and sometimes even crucial—they should be used sparingly and only when absolutely necessary.

Make sure the demo is actually a demo. The goal of the demo is to show what was accomplished and get feedback, not to impress everyone with your vast technical knowledge.

Remember, you're presenting to a diverse group of people with different backgrounds and expertise. It's important to keep that in mind when preparing the demo.

3. Do Your Best to Avoid Surprises

Bad things happen, and demos are not immune. To avoid surprises, the presenter should run through the demo before the actual meeting. This will help identify any potential issues and give the presenter a chance to come up with a plan B in case something does go wrong.

Now, one thing that often gets overlooked is the fact that sprint reviews are usually done via video conferencing. The video conferencing software may interfere with the demo in many ways, such as introducing lag, slowing down or even crashing the presenter's computer, increasing network latency, cutting off the presenter's audio, or freezing the video.

To avoid these issues, I'd recommend doing a test run of the demo with the same video conferencing software that will be used for the actual meeting.

4. Don't Turn the Sprint Review into a Blame Game

Sprint reviews should be a positive experience where everyone comes together to celebrate the team's accomplishments.

If something goes wrong during the sprint, the focus should be on finding a solution, not on assigning blame. The whole team should work together to come up with a plan to avoid the same mistakes in the future.

Sprint reviews should not be used as a platform to blame each other or to point fingers. That's the worst possible outcome, and it should be avoided at all costs because it fosters a toxic environment of fear and mistrust.

5. Be Prepared for Questions and Feedback

After the demo, there is usually time for a brief Q&A session. This is where stakeholders and the product owner can ask questions and provide feedback. It's important to be prepared for this part of the meeting.

We once had a stakeholder ask why we chose to use a particular technology, and the presenter didn't have a good answer. It was an uncomfortable (awkward) moment, to say the least.

Don't get me wrong: the Q&A session is not about grilling the team with tough questions. It's about open communication and collaboration. But it's important to be prepared for questions and feedback, nonetheless.

The team should come up with some good answers ahead of time, and presenters should be prepared to defend the team's choices.

6. Practice the Art of Demoing

Preparing a demo is not easy. Like any other skill, it takes time and practice to get it right.

If you're just starting out, I'd recommend finding a demo buddy. Find someone who can help you prepare for the demo, give you feedback, and practice with you. This will make a world of difference.

Also, preparing a script or a bulleted list of points to cover during the demo can be helpful. This will ensure that you don't forget anything important, and it will help keep the demo on track.

7. Have Fun!

Finally, don't forget to have fun with it. The sprint review is a chance to show off your team's accomplishments and get everyone excited about the project. Enjoy it!

Sprint reviews—when done right—can be a great way to get everyone on the same page and collect feedback. But it takes time and practice to master the art of the perfect demo.

The most important thing is to keep the audience in mind and focus on what's relevant to them.

3.11 The Sprint Retrospective: Reflecting on the Past

The sprint retrospective is a meeting that's held at the end of every sprint. It's an opportunity for the team to reflect on the past sprint. The goal is to identify what went well, what could be improved, and what actions should be taken to make improvements in the future.

The Scrum Master facilitates the meeting, and all members of the development team should participate.

Neither product owners nor stakeholders should attend the retrospective, as it's meant to be a safe space for the team to openly share their thoughts and ideas.

I'm not sure whether the retrospective or the daily standup felt more awkward when I first started out with Scrum. But I can tell you that, over time, the retrospective has become one of my favorite meetings. I wouldn't even call it a meeting anymore, but more of a chance for the team to get together and have an open, honest conversation about what's working and what's not. It's a chance to come up with creative solutions to problems. And it's a chance to build trust and collaboration within the team.

Okay, let's get to the practical stuff: How to run a sprint retrospective. I'll share with you the approach that I've found to be the most effective.

Let's say you just completed a sprint. The sprint review went relatively well, but there were a few hiccups here and there. The team could not quite meet the sprint goal, and there were a couple of bugs that interfered with the demo. In addition, one of the stakeholders asked a tough question that the team was not prepared to answer.

At the beginning of the retrospective, everyone has a chance to share their thoughts on the past sprint. This can be done in a variety of ways, such as through a round-robin discussion or brainstorming. Here's what worked for us.

First, we went around the room, and each person shared one thing that they thought went well in the past sprint. For instance:

→ "I thought the team did a great job of communicating with each other"
→ "I was really happy with how we handled the unexpected bug before the demo."

It's crucial to start the retrospective on a positive note. This sets the tone for the rest of the meeting and allows everyone to share their thoughts openly.

The speaker should be brief (one to two minutes at most) and should avoid getting into too much detail. Interruptions are not allowed at this stage.

The Scrum Master or another designated team member should take notes during this part of the retrospective so that all ideas are captured. We usually used a whiteboard for this or sometimes a text editor that was shared using a projector. Bottom line: make sure everyone can see the ideas that are being generated.

After everyone has had a chance to share their thoughts on what went well, it's time to move on to the next topic: what could be improved. Just like before, each person takes a turn sharing their thoughts:

- "I think we could have done a better job of estimating tasks that involve research."
- "We need to improve our network layer to prevent sporadic bugs like the ones we saw during the demo."

These ideas should be captured in the same way as the positive ones. Again, the speaker should be brief and to the point.

After that, we brainstormed potential actions that we could take to improve the situation:

- "Before each review, we should test run the demo while Zoom is open so we can see any potential glitches."
- "In the future, let's make sure to set aside time for research when we're estimating tasks."
- "Improve the network layer's unit test coverage to help us catch bugs before they make it to production."

Finally, the team decides on the top few actions they want to focus on for the next sprint. This is important because it helps the team focus on a few key areas rather than trying to tackle everything at once. These items are then added to the next sprint's backlog. What happens with the rest of the ideas that were generated? They're added to a "parking lot," which is essentially a list of potential improvements that the team can come back to at a later time.

So, that's how we've been running our retrospectives. This simple process has helped us improve our sprints and learn from our mistakes. I hope it does the same for you!

3.12 Documentation in Scrum - Misconceptions and Best Practices

One of the most common questions I get is, "How much documentation should we have in Scrum?" There seems to be a lot of confusion around this topic, so I wanted to clear things up.

«Working software over comprehensive documentation» - this is probably the most misinterpreted principle in the Agile Manifesto. Some people see it as a call to completely eliminate documentation, while others view it as a green light to produce low-quality documentation.

"Agile means no documentation," "We don't need documentation because we're Agile," and variations thereof are unfortunately common misconceptions. These statements couldn't be further from the truth!

Let's start by analyzing the principle itself:

> "Working software over comprehensive documentation."

The first thing to note is that the word "comprehensive" is used, not "any." The Manifesto is not advocating for the complete elimination of documentation but rather for a shift in priorities.

We've already talked about the traditional Waterfall development model, in which documentation is produced before any code is written. The problem with this approach is that we spend a lot of time and effort creating documentation, such as requirements, specifications, and design documents, before writing a single line of code. As you may recall, this often leads to redoing work because the documentation is based on assumptions that may no longer be valid by the time we start building the software. The requirements are carved in stone, so to speak, and any changes that need to be made down the road are very costly.

The Agile Manifesto advocates a different approach in which working software gets prioritized over comprehensive documentation. In other words, the focus should be on building software that meets the customer's needs, not generating mountains of text.

Of course, this doesn't mean that we should eliminate documentation altogether. How could we build software without documenting requirements, architecture, progress, and so on? The difference is that in Agile, we produce documentation as needed and with the level of detail appropriate for the current situation—and not all at once at the beginning of the project. In other words, documentation should be created when it adds value to the product.

Some of our projects required extensive design documents with detailed UML diagrams, performance and security analysis, usability testing plans, and so on. Other projects required only a few high-level documents, such as product and sprint backlogs.

The idea is to produce only the documentation that's needed and when it's needed.

Now, I can't give you a recipe for how much documentation you need because it really varies from project to project and from team to team. But I can give you a few general guidelines:

If the documentation is not being used, it's probably not needed. So, if you find yourself with a lot of documentation that's gathering dust, it may be time to rethink your approach. Say you've created a thorough design document, but no one ever looks at it. Yet, the project progresses just fine without it. In this case, it may make sense to eliminate or simplify the document to the bare minimum.

On the contrary, if your team constantly asks the same questions about implementation details, it's probably time to collect all the technical decisions and put them in a design document that everyone can use as a reference.

Your documents should be concise, well-organized, and easy to read and understand. In my experience, ancient corporate templates or old habits are usually to blame when documentation is excessively long and convoluted. You can't just reuse the same 100-page requirements document from the 90s for your new project unless you want to put everyone to sleep.

And here's my final piece of advice: review and update your documents regularly. A document that's not kept up-to-date is just as bad as no document at all.

So, these are just a few general guidelines to keep in mind when it comes to documentation.

To Sum Up:

Create documentation only when needed, and ensure it provides value to the project.

We arrived to the end of this module. I hope you now have a better understanding of Scrum and how it can help your team be more productive. In the following chapter, we'll look at how Kanban can help you visualize and manage your work.

4. KANBAN

4.1 Introducing Kanban

Kanban will probably sound familiar to you if you've ever used a task management tool like Trello, Asana, or any other board-based system. That's because Kanban is all about visualizing your work so that you can see what needs to be done and when.

The word Kanban means "signboard" or "billboard" in Japanese. This visual management system was created in 1940 by Taiichi Ohno, an Industrial Engineer at Toyota Motor Corporation, to increase the efficiency of the company's manufacturing processes.

The basic idea behind Kanban is to make the workflow visible so that bottlenecks and inefficiencies can be identified and addressed. If inventory levels for a particular part fall below a specified threshold, a Kanban card is used to signal the need to produce more of that part. The Kanban card is basically a message from the consumer to the supplier that more parts are needed.

Unlike the push-based system of the traditional assembly lines—meaning that parts are produced even if there's no demand for them—Kanban is a pull-based system, which means that parts are produced based on actual demand. This approach eliminates overproduction and optimizes inventory levels, which leads to cost savings and increased efficiency.

In the early 2000s, a software engineer named David Anderson began adapting the Kanban manufacturing process for software development. His efforts led to the creation of the Kanban Method, which is now used by software development teams all over the world.

Kanban is a flexible system that can be adapted to fit the needs of any organization. In its simplest form, it consists of three steps:

→ Visualizing the work
  This means creating a visual representation of your work, e.g. a Kanban board.

→ Limiting work in progress
  By setting limits on the number of tasks that can be in progress at any given time, you can avoid bottlenecks and ensure that tasks are completed on time.

→ Improving continuously
  By constantly evaluating and improving the way you work, you can identify bottlenecks and inefficiencies and find ways to eliminate them.

The idea behind Kanban is pretty simple: you have a board with columns representing the different stages of your workflow and cards depicting the tasks that need to be completed. In a typical setup, you might have columns for "To Do," "In Progress," and "Done."

Each developer is assigned a task, which they move from one stage to the next as they make progress.

Let's say you start working on a task. Initially, it would be in the "To Do" column. Once you start working on it, move it to the "In Progress" column. When you're finished, move it to the "Done" column.

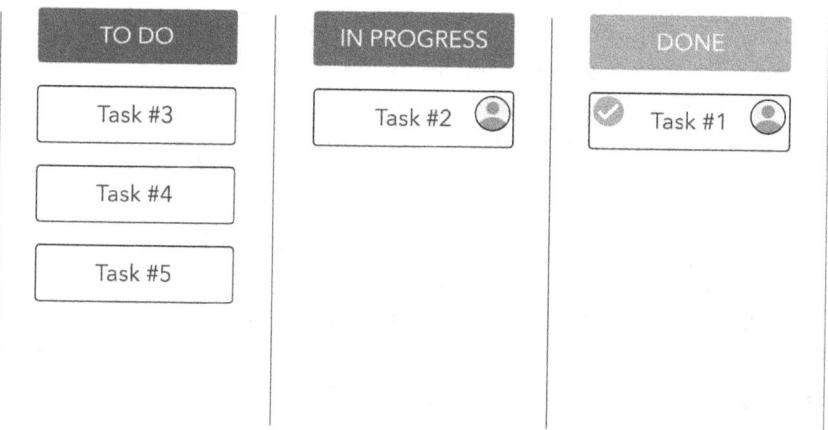

It's that simple. Imagine how valuable this simplicity can be when you have a team of developers working on a project with dozens or even hundreds of tasks.

Kanban makes it easy to see what needs to be done, track the progress of each task, and help identify and address bottlenecks in the workflow.

It's also worth noting that Kanban boards can be as straightforward or as complex as you need them to be. For instance, you might add additional columns for different stages of work, such as "In Review," "QA," or "Deploy."

The point is that the Kanban system is flexible, and you can customize it to fit the needs of your team.

Let's take another example: Say you're working on a project with a large team. You can create Kanban boards for each developer or workflow stage.

You might have one board for the front-end and another for the back-end development tasks. Or you might create one board for the design tasks and another for the coding tasks—the possibilities are endless. The important thing is that Kanban can be adapted to fit the needs of almost any organization or project, no matter how big or small.

I love that Kanban is a simple system that can be used to great effect. It's easy to learn and implement but also extremely powerful.

Story Time

Let me tell you a personal story that illustrates this point. A few years ago, I was working on a project with a team of developers, and we were having some trouble staying organized. We had too many tasks in progress, and things were falling through the cracks. At the same time, we were also using many different tools, and it wasn't easy to get a clear overview of what needed to be done.

I accidentally saw another team using a Kanban board and asked them about it. They explained the system to me, and I was sold.

We started using Kanban, and it changed everything. Suddenly, we had a clear overview of all the tasks that needed to be completed. We could see at a glance what was in progress and what needed to be done next. It was like a "gamified" version of the Excel spreadsheet we had been using that was way more fun and effective.

Since then, I've been a huge advocate of Kanban. I've used it on countless projects, and it never fails to impress me. If you're looking for a way to increase your team's productivity, I highly recommend giving Kanban a try.

In the following section, we'll look at how to set up a Kanban board using Asana.

4.2 Setting up a Kanban Board in Asana

In this lecture, we're going to take a look at how to set up a Kanban board using Asana.

If you're unfamiliar with Asana, it's a web-based project management tool incredibly popular with Agile teams.

To get started, go to https://asana.com and create an account. The registration process is quick and easy, and you should be up and running in no time.

Once you're logged in, create a new project, and name it "Kanban Demo." Next, choose "Board" as the Default view type and hit Continue.

At this point, you can start adding team members to your project, but for now, we'll just add tasks.

So, leave the default "Start adding tasks" selection and click the "Go to project" button.

As you can see, Asana has automatically created a few columns for us: "To do," "In progress," and "Complete."

These are the default columns, but you can rename or delete them. If you need additional columns, click the "Add section" button on the right side of the screen.

For this demo, we'll stick with the default columns.

Now, let's create the first tasks. Go ahead, and add a task called "Create Kanban Board" in the "To do" column.

There are a few different ways to create the first task:

- You can rename the task that was automatically created for you
- You can add a new task by clicking the "Add task" button
- Alternatively, you can click the '+' sign next to the "To do" column.

Let's edit the automatically created task named "Task 1." Click the task and change the name to "Create Kanban Board."

Asana lets you set the assignee—the person responsible for completing the task—as well as the due date and a description.

You can create subtasks in case the task is too big to be completed in one go, and you can also add attachments and perform other actions.

When you're done, hit the Esc button or just click outside the task to save your changes.

Now let's create a few more tasks:

- Add a task called "Add tasks to Kanban Board."
- Create another one called "Demonstrate how Kanban Board works."

To move a task to a different column, click and drag it to the desired location. You can also click on the task and change the column type in the detail view by clicking on the drop-down menu.

We have already completed the first task, "Create Kanban board." So, move it to the "Complete" column using the drag-and-drop method.

We're in the process of adding tasks to the Kanban board. Thus, move that task to the "In progress" column.

How about the "Demonstrate how Kanban Board works" task? Well, that one is "In progress" because we're currently working on it.

As you can see, setting up a Kanban board in Asana is a breeze. Other web-based Kanban tools are also available, such as Trello and Jira. You can use a physical Kanban board, which works great if your team is based in the same location. However, a web-based tool is probably the best option for remote teams.

Next, we'll look at how to use Kanban to increase your team's productivity.

4.3 The Importance of Limiting Work In Progress

Kanban aims to help teams work more efficiently and prevent them from becoming overwhelmed by too much work. In order to do that, it's important to limit the amount of work in progress (WIP).

Work in progress is defined as any task that has been started but not yet completed.

For example, if you're working on a task and you've made some progress, but you're not done yet, that task is considered "work in progress." If you start working on another task before finishing the first one, you have two tasks in progress.

Too much work in progress can lead to context switching, which is when you have to switch your focus from one task to another. This can be very disruptive and lead to lost productivity.

Now, if you add more and more tasks to your plate, the amount of context switching will increase, and you'll eventually reach a point where you're completely overwhelmed and unable to get anything done.

And if your entire team is in the same situation, the whole team will be ineffective. This leads to frustration and, ultimately, to project failures.

So, how do you avoid this? By limiting the amount of work in progress.

There are a few different ways to do this. One way is to set a limit on the number of tasks that can be in progress at any given time. Say, each person can only have three tasks in progress at a time. The Kanban board can be used to enforce this limit by making it visually obvious when someone has too many tasks in progress. For example, the "In Progress" column would turn red when someone has more than three tasks in it.

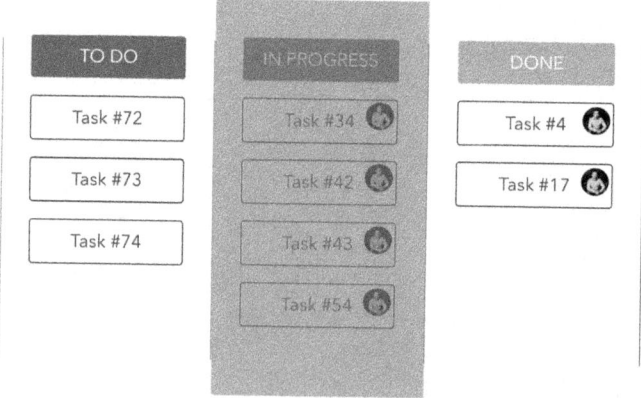

Another way to limit work in progress is to restrict the maximum number of tasks in the Kanban board columns. This "scarcity" of tasks will force people to finish their work before starting new tasks.

Here's what worked for my team: we used a whiteboard to visualize our Kanban board. It had five columns: "To Do," "In Progress," "Under Review," "Testing," and "Completed."

"Under Review" means that the task is peer-reviewed by someone else on the team. In my opinion, code reviews are crucial, regardless of whether you're using Agile, Kanban, or any other methodology. "Testing" is self-explanatory. Once the code is reviewed and approved, it needs to be tested before it can be considered "done."

We also had a rule that no one could start working on a new task until they had finished their current task and moved it to the next column. In other words, the number of tasks in the "In Progress" column was limited to the team's size. So, if we had five people on the team, we could have a maximum of five tasks in progress. This really helped us focus and get our work done more efficiently.

There are many other ways to limit work in progress. What's important is to find a method that works for your team and stick to it.

Limiting work in progress is a critical concept in Kanban, and it's one of the things that makes it so effective.

4.4 Creating a Cumulative Flow Diagram in Excel

In the previous lecture, we talked about limiting work in progress to increase productivity. Now, while Kanban boards are a great way to visualize work in progress, they don't tell you much about the overall progress of your project over time.

To get this kind of information, you can use a cumulative flow diagram (in short, CFD). The aim of this chart is to provide insight into past and current performance and make it easier to identify bottlenecks or other problems that are slowing down your team. A cumulative flow diagram can visualize three key metrics:

- The cycle time—the time it takes for a particular task to move from one stage of the workflow to another.

- The throughput, which measures the average number of tasks completed by the team in a given time period.

- The work in progress, which shows the number of tasks that are currently in progress.

These metrics can help you understand how well your team is performing and what steps you can take to improve productivity.

There are many different ways to create a cumulative flow diagram. You can use a dedicated software tool, but any spreadsheet or diagramming software will work as well.

Let me show you how to create a CFD for your team. I'll walk you through the basics of creating the diagram in Excel.

First, download the starter excel file available at https://github.com/nyisztor/cfd.

Next, open the starter file in Excel. You should see a table with columns for "To Do," "In Progress," "Testing," "Completed," and an additional "Date" column. We need the latter to track the progress over time, which is essential for creating a cumulative flow diagram.

I've pre-populated the table with dates and the total number of tasks at each stage for that particular day.

To create the cumulative flow diagram, select the data by holding down the **Shift** key and dragging your mouse over the table. Now, click the Insert menu and select **"Stacked Area"** from the **Recommended Charts** dropdown.

A new chart will appear. Position it below the table and scale it to fit the data by dragging and resizing the chart.

Each colored band represents a stage of the workflow, and its height is proportional to the total number of tasks in that stage. Next, we want to reverse the bands on this chart so that the last stage appears at the bottom and the first stage at the top. This order is important for the chart to work as a cumulative flow diagram.

To reverse the bands, right-click inside the chart and choose "Select Data." Reverse the order of the legend entries using the arrow keys. "To Do" moves to the bottom, "Completed" moves to the top, and "Testing" should come right after "Completed."

To finish up, change the Chart Title to Cumulative Flow Diagram.

And we're done with this chart! But how can we use it to improve our productivity? I'll show you how to interpret the cumulative flow diagram in the next section.

4.5 How to Interpret a CFD?

Now that we have a cumulative flow diagram, it's time to figure out what it can tell us about our team's performance.

Let's first take a closer look at the chart.

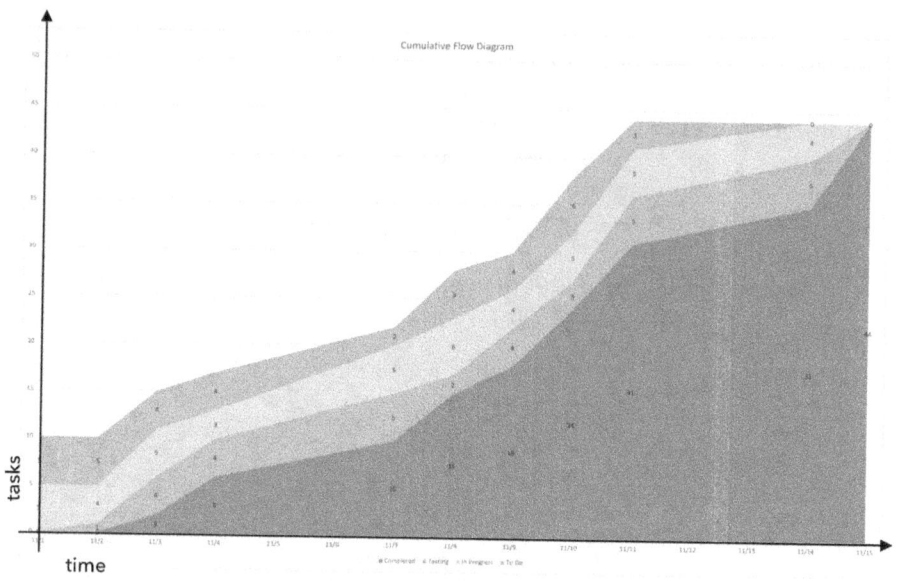

The horizontal or x-axis represents the time frame for which the chart captures data. The time unit can be either days, weeks, months, or whatever makes sense for your team. I used days in this example.

The vertical y-axis represents the total number of tasks in the workflow at a given point in time.

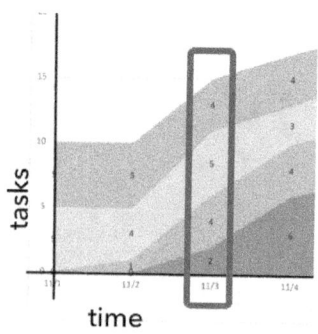

For instance, we can see that on November 3rd, there were a total of 15 tasks in the workflow. The height of each band represents the number of tasks for that stage on that day.

To see the actual counts, right-click on each band and select "Add Data Labels." Now it's even easier to read the data.

The number of tasks in a particular stage will vary as tasks move to the next stage and new tasks are added. Thus, the height of each band is constantly changing over time. They can shrink, grow, or even flatline if the task count on a particular day becomes zero.

The only exception is the "Completed" stage, which will keep rising as the team completes more and more tasks. The slope of this band reveals how fast the team completes tasks: the steeper the slope, the faster the team finishes tasks.

Checking how the chart progresses over time provides clues on your team's overall productivity.

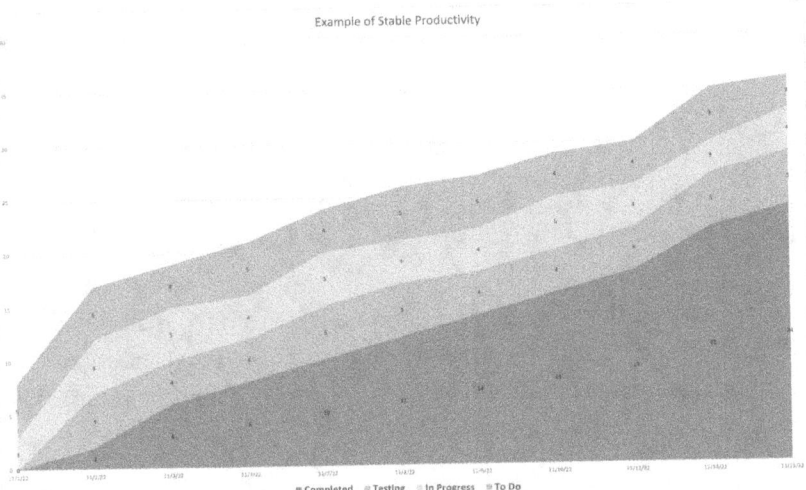
Example of Stable Productivity

If the bands are progressing in parallel, the team's productivity is stable: the number of tasks entering the workflow is roughly equal to the number leaving it, and the total count of tasks in each stage is constant.

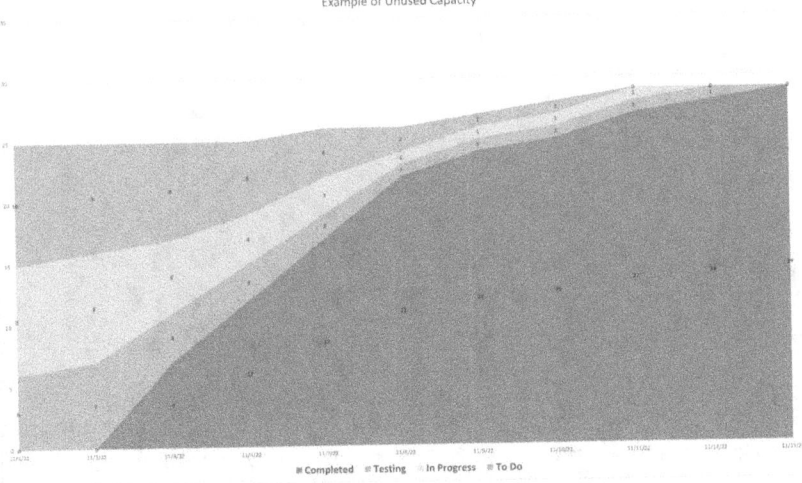
Example of Unused Capacity

Rapidly narrowing bands indicate that your team is burning through tasks at a faster rate than they're entering the workflow. That means you have unused capacity and can likely take on more work.

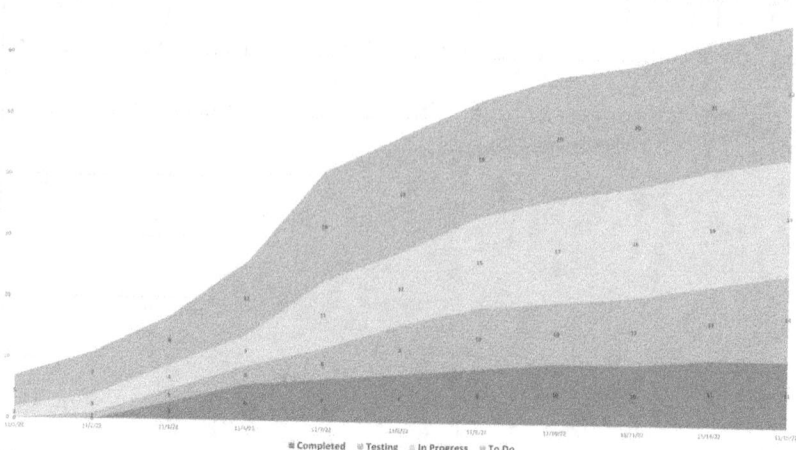

Example of Team Falling Behind

On the contrary, if you're noticing widening bands on your chart, it's a sign that your team is falling behind. If this trend persists over time, you'll need to find ways to improve your team's productivity or consider hiring new developers to pick up the slack.

Bear in mind that none of the bands should ever go down. That would mean that tasks disappear from the workflow—an impossible scenario. If you're seeing this happen, there is likely a data entry error in your CFD.

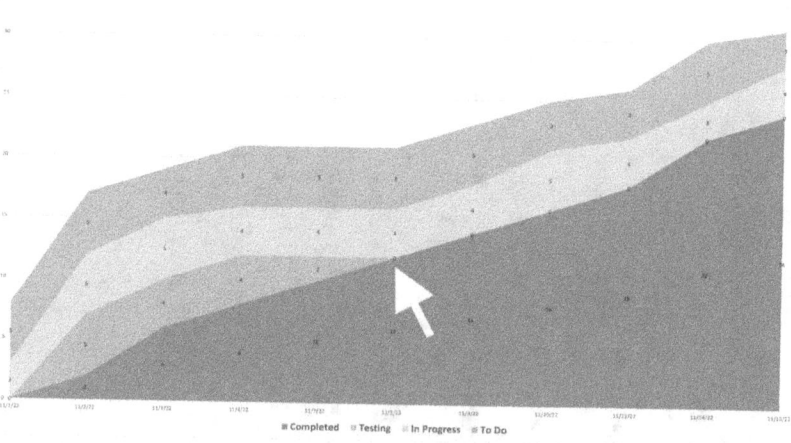

Example of Unused Testing Stage

However, some bands may be flatlining. This means that no new tasks are being added to the workflow. Thus, that specific stage is not being used for some reason.

Now, let's see how the cumulative flow diagram can help us measure the approximate average cycle time, throughput rate, and work in progress—the three crucial metrics I mentioned in the previous section.

To measure the approximate average cycle time, we need to measure the time from when a task enters the workflow until it exits it. Thus, we'll look at the horizontal distance between the top of the first stage—our "To Do" queue—and the top of the "Completed" stage.

For instance, here's the top of the "To Do" band on November 3. Now let's look at the top of the "Completed" stage by drawing a horizontal line between the two points. We can repeat this process for each point in time. As you can see, the distance between the top of the "To Do" and the top of the "Completed" band is around three days in each case, which indicates an average cycle time of roughly three days.

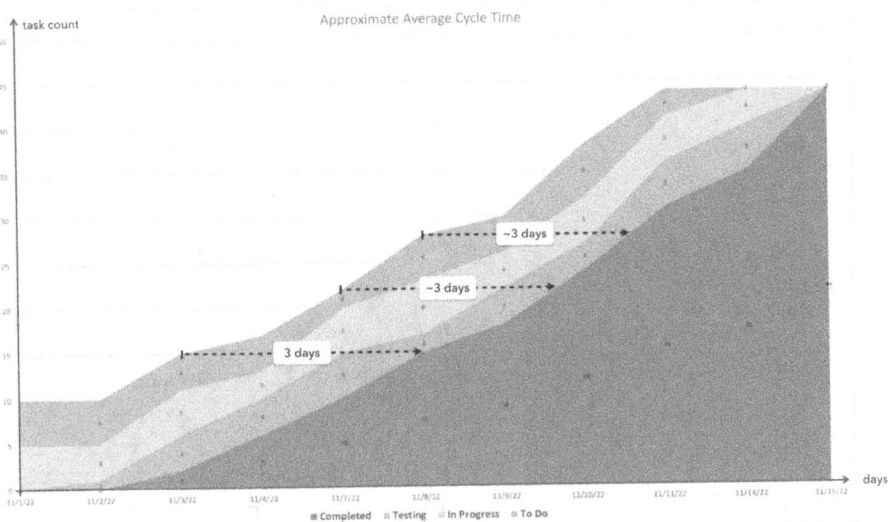

To measure work in progress (or how many tasks are currently active or queued for processing), we'll look at the vertical distance between the top of the "To Do" band and the top of the "Completed" band. Take the vertical distance on November 3, for example. We can see that the total number of tasks in the "To Do," "In Progress," and "Testing" stages totals 13. So, the work in progress is roughly 13 tasks at that point in time.

And finally, to measure the throughput rate—which gives us the average number of tasks completed by the team in a day—we'll need to calculate the slope of the "Completed" line between any two points on the chart.

How to calculate the slope? Simple: it's the rise divided by the run. The rise is the vertical change between two points, and the run is the horizontal change between them.

So, the rise of the "Completed" line between November 3 and 8 is the vertical distance between the top of the "Completed" band on November 3 and the top of the "Completed" band on November 8, which is 13.

And the run is the horizontal distance between these points, which is three. So, the slope is 13/3, which is approximately 4.3. So, we can conclude that the team's average throughput rate is roughly four tasks per day.

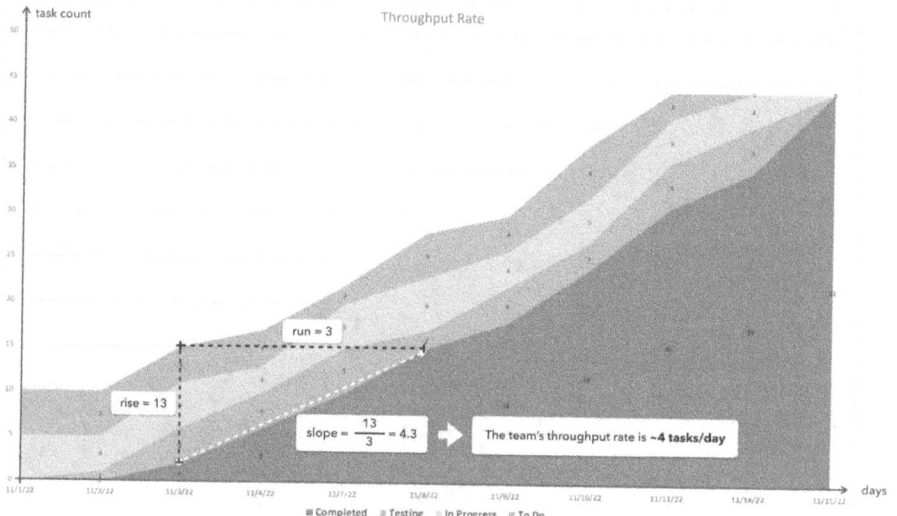

As you can see, interpreting the Kanban Cumulative Flow Diagram is not rocket science. By studying the trends and patterns over time, we can easily monitor progress, identify bottlenecks and make informed decisions about how to keep our projects on track.

4.6 Goodbye!

Congratulations! You made it!

This knowledge is going to be invaluable as you continue your journey in the world of software development. Whether you're looking to implement Agile, Scrum, or Kanban practices in your own development team, you now have the foundation necessary to succeed.

I'd encourage you to revisit this book now and again. You will discover new ideas, and you'll see things differently the next time around.

The video course version of this book is available on Udemy at https://www.udemy.com/course/agile-scrum-and-kanban-foundations. The course comes with quizzes, downloadable resources, and bonus content.

Best of luck with your projects!

Best Regards,
Karoly Nyisztor

www.ingramcontent.com/pod-product-compliance
Lightning Source LLC
Chambersburg PA
CBHW070318220526
45465CB00004B/1891